Meat and Protein

Ann Thomas

An Imprint of Chelsea House Publishers
A Haights Cross Communications Company
Philadelphia

Chelsea Clubhouse
1974 Sproul Road, Suite 400
Broomall, PA 19008-0914

The Chelsea House world wide web address is www.chelseahouse.com

Library of Congress Cataloging-in-Publication Data

Thomas, Ann, 1953-
    Meat and protein / by Ann Thomas.
    p. cm. — (Food)

    Includes index.
    Summary: Presents information on the meat and protein food group, describing various sources of protein and how they are prepared.

    ISBN 0-7910-6978-8
    1. Meat—Juvenile literature. [1. Proteins. 2. Meat. 3. Nutrition.] I. Title. II. Food (Philadelphia, Pa.)
    TX373 .T46 2003
    641.3'6—dc21

                                                                    2002000031

First published in 1998 by
MACMILLAN EDUCATION AUSTRALIA PTY LTD
627 Chapel Street, South Yarra, Australia, 3141

Copyright © Ann Thomas 1998
Copyright in photographs © individual photographers as credited

Text design by Polar Design
Cover design by Linda Forss
Illustrations © Anthony Pike

Printed in China

Acknowledgements
Cover: Photolibrary.com

Australian Picture Library, pp. 5 ©Bob Walden, 14 ©Oliver Strewe, 24; Coo-ee Picture Library, pp. 10, 15, 20; Corbis, p. 27; Getty Images, p. 26; Great Southern Stock, pp. 8, 18, 29; HORIZON Photo Library, pp. 11, 22, 23; Dale Mann/Retrospect, p. 30; Photolibrary.com, pp. 6 ©Jenny Mills, 16, 17 ©Ken Stepnell, 21 ©Peter Fogg, 25 ©Simon Jauncey, 28 ©Seth Joel/SOL/Stock Photos, pp. 4 ©Benaji, 9 ©Tony Moran, 13 ©Paul Steel, 19 ©Rick Altman; U.S. Department of Agriculture (USDA), p. 7.

While every care has been taken to trace and acknowledge copyright, the publisher tenders their apologies for any accidental infringement where copyright has proved untraceable.

Contents

Why Do We Need Food?

We need food to keep us healthy. All living things need food and water to survive.

Hot dogs are made from ground meat.

Some birds eat berries or drink nectar from plants.

There are many kinds of food to eat. People, animals, and plants need different types of food.

What Do We Need to Eat?

Foods can be put into groups. Some groups give us **vitamins** or **minerals**. Some groups give us **proteins** or **carbohydrates**. We need these **nutrients** to keep us healthy.

We need to eat a variety of foods.

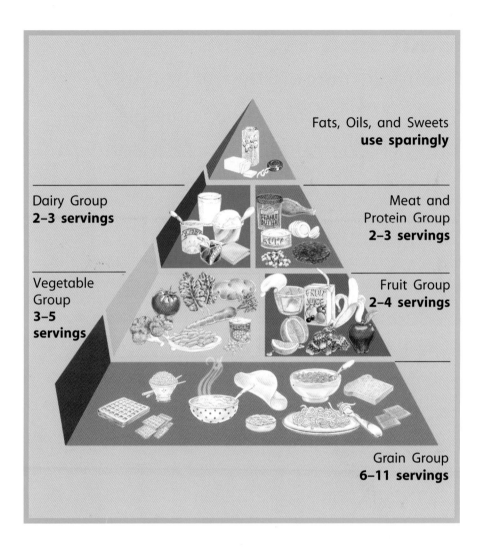

Fats, Oils, and Sweets
use sparingly

Dairy Group
2–3 servings

Meat and
Protein Group
2–3 servings

Vegetable
Group
**3–5
servings**

Fruit Group
2–4 servings

Grain Group
6–11 servings

The food guide pyramid shows us the
food groups. We should eat the least
from groups at the top. We should eat
the most from groups at the bottom.

Meat and Protein

One food group is the meat and protein group. Meat comes from animals. Beef, lamb, chicken, and pork are meat. Fish and seafood are meat, too.

beef   chicken   lamb

pork

shrimp   sardines   crayfish   mussels

oyster   scallop   crab

All types of meat are rich in protein.
Protein gives us energy. Foods like eggs,
dried beans, dried peas, and nuts also
give us protein.

Types of Meat

Long ago, people hunted wild animals for meat. Most of the meat we eat today comes from farm animals. Beef comes from cattle raised on ranches or farms.

Pork comes from pigs raised on pig
farms. Lamb comes from sheep raised
on sheep farms. Chicken farms and
turkey farms provide meat, too.
Chickens are also raised for their eggs.

Preparing Meat

Workers at meat-packing plants prepare the meat we eat. They keep the meat in refrigerated areas so it stays fresh.

Inspectors make sure the meat is clean and safe to eat.

Workers clean the meat. They hang large pieces of meat from hooks. **Inspectors** check the meat before it is sold to stores.

Buying Meat

Butchers work at supermarkets and meat stores. They trim the meat carefully to remove excess fat. Then they cut meat into smaller pieces and put it in packages.

People can buy many types of meat. They may want beef steak, pork chops, or chicken breasts. Sausages and hot dogs come from ground up meat.

Fishing

Fishers catch fish on poles, in nets, and in traps. They store fish on ice or in cold bins. Fishers usually sell their catch at a fish market.

Fishers empty their catch from the large fishing net.

Trout is being raised at this fish farm.

Fish farmers raise fish such as trout, catfish, and salmon in large ponds. Oysters, lobsters, and other seafood are also grown on fish farms.

Preparing and Buying Fish

Large fish markets sell fish to supermarkets and restaurants. Workers sometimes remove the scales and bones from the fish. Fish pieces prepared without bones are called fillets.

A fish market's business usually takes place early in the morning.

People buy whole fish or fillets at
supermarkets. People sometimes
buy fresh oysters, shrimp, and other
shellfish that are still in their shells.

Unusual Kinds of Meat

Many animals can be used for meat. Buffalo, ostrich, or crocodile meat is sometimes sold in stores and restaurants.

Crocodile farms raise crocodiles for the meat market.

Cooked snails are called escargot. This is a popular French dish.

Some people eat frogs' legs, snails, or
pâté made from duck or chicken liver.
Pâté is spread on bread or crackers.

Eating Meat

Beef, lamb, pork, chicken, and other meats can be prepared many ways. They can be roasted, baked, grilled, fried, dried, or smoked.

Turkish shish kebabs are usually spicy.

Cubes of meat cooked on **skewers** are called shish kebabs. Meat is also enjoyed in soups, stews, and mixed with other foods.

Fish and other seafood are also prepared in many ways. They can be grilled, baked, fried, boiled, steamed, or smoked.

Deep-fried fish served with hot french fries is a popular meal.

Eggs

Eggs are another source of protein. Most people eat eggs from chickens. Some people eat duck eggs or eggs from wild birds.

26

People eat fried, scrambled, and boiled eggs. They make **omelettes** and other egg dishes. They also use eggs in batter for cookies and cakes.

Other Sources of Protein

Some people do not eat meat. They are vegetarians. Some people do not eat meat or eggs, and they do not drink milk. They are vegans. Many vegetarians and vegans eat dried beans, dried peas, and nuts to get their protein.

Dried beans and dried peas have protein.

Many foods can be made without animal products.

Soy milk and tofu are made from soybeans. Tofu can be cooked like meat or made into other foods. Veggie burgers are made from vegetables and grains.

The Meat and Protein Group

We should eat two to three servings from the meat and protein group each day.

tofu   eggs   meat

dried beans

Glossary

butcher   a person who cuts up and sells meat

carbohydrate   an element found in certain foods that gives us energy when eaten; bananas, corn, potatoes, rice, and bread are high in carbohydrates.

fisher   a person who catches fish for a job or for fun

inspector   a person who checks or examines things for safety

mineral   an element from earth that is found in certain foods; iron and calcium are minerals; we need small amounts of some minerals to stay healthy.

nutrient   an element in food that living things need to stay healthy; proteins, minerals, and vitamins are nutrients.

omelette   beaten eggs that are not stirred while cooking; the cooked eggs are folded in half; people sometimes put cheese, meat, or vegetables inside the omelette.

protein   an element found in certain foods that gives us energy when eaten; eggs, meat, cheese, and milk are high in protein.

skewer   a pointed stick of metal or wood used to hold pieces of meat or vegetables while cooking

vitamin   an element found in certain foods; Vitamin C is found in oranges and other foods; we need to eat foods with vitamins to stay healthy.

Index